THANK YOU FOR YOUR CONTINUED SUPPORT.

Published By My Littl' Workshop

To purchase additional
copies of this book, go to:
www.MyLittlWorkshop.com

My Littl' Workshop supports Sista Foundation Inc. a nonprofit which empowers young women and mentoring young men, and communities at large through workshops and social events.

This Notebook Belongs to

____

____

Subject: ________________________

Grade: _________________________

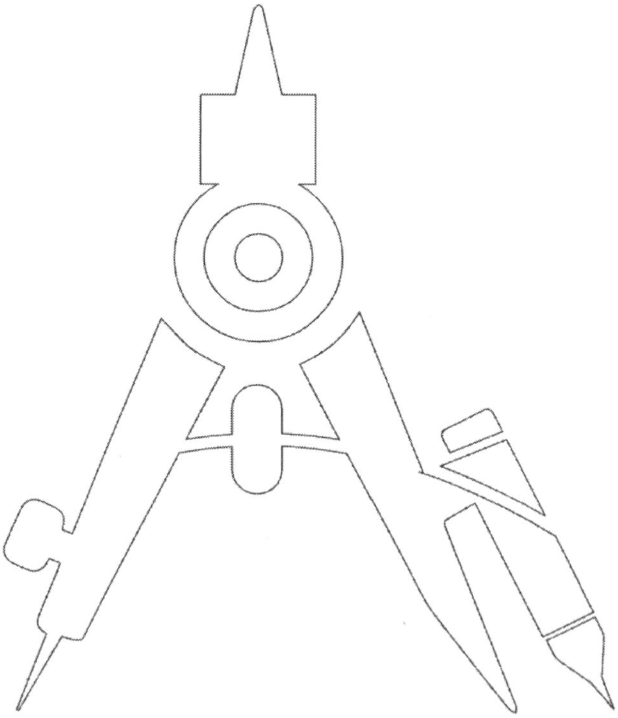

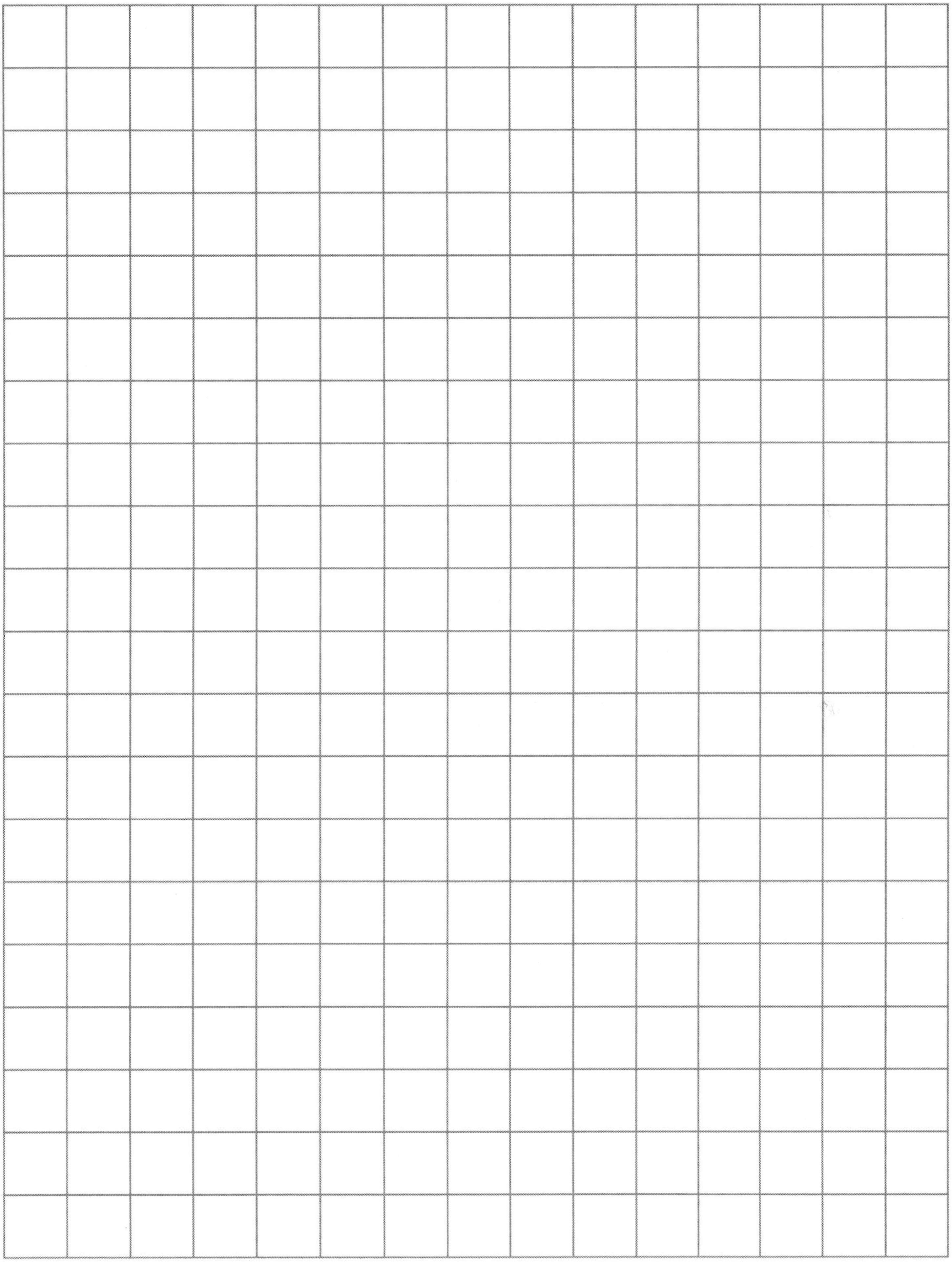

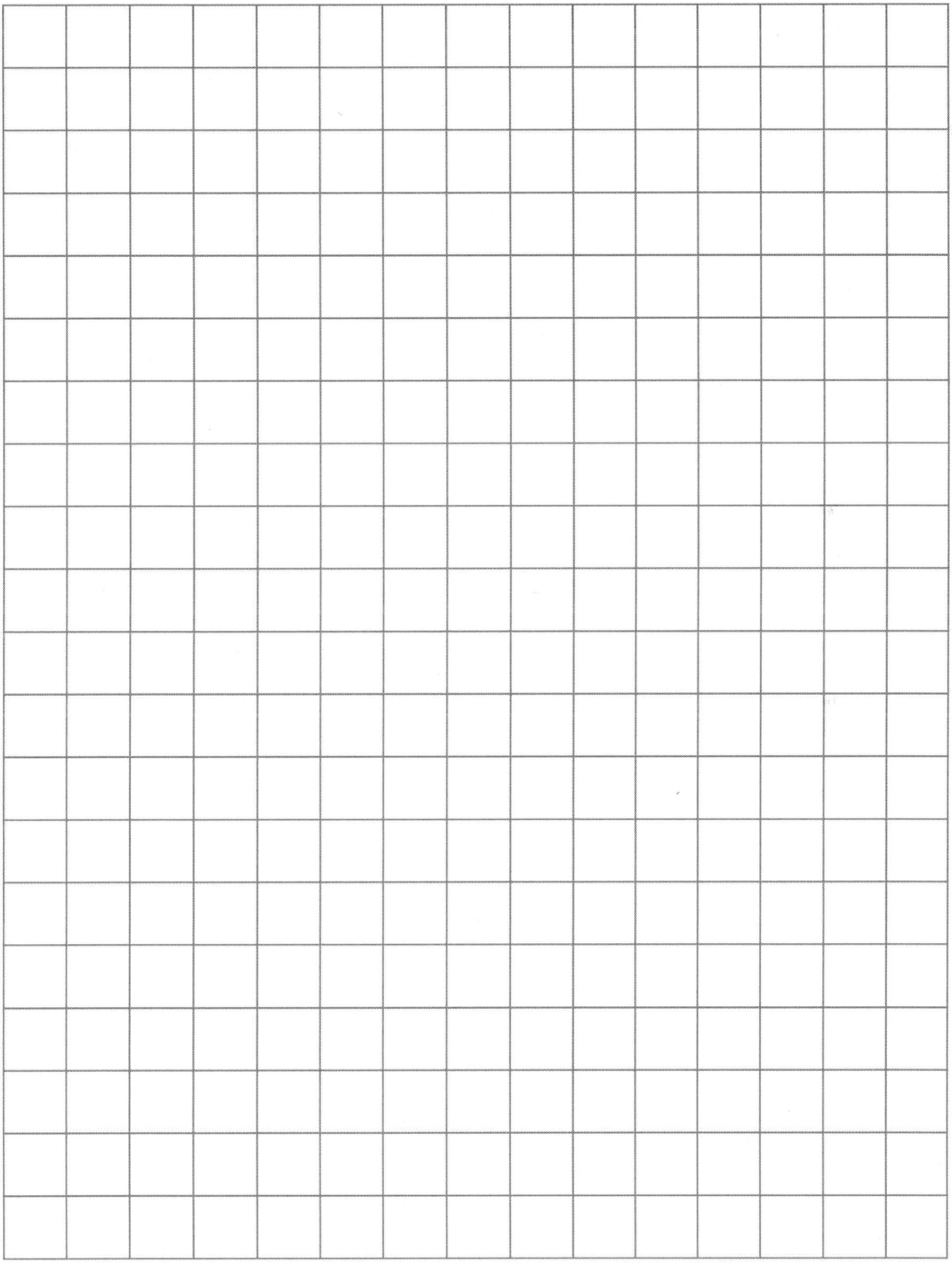

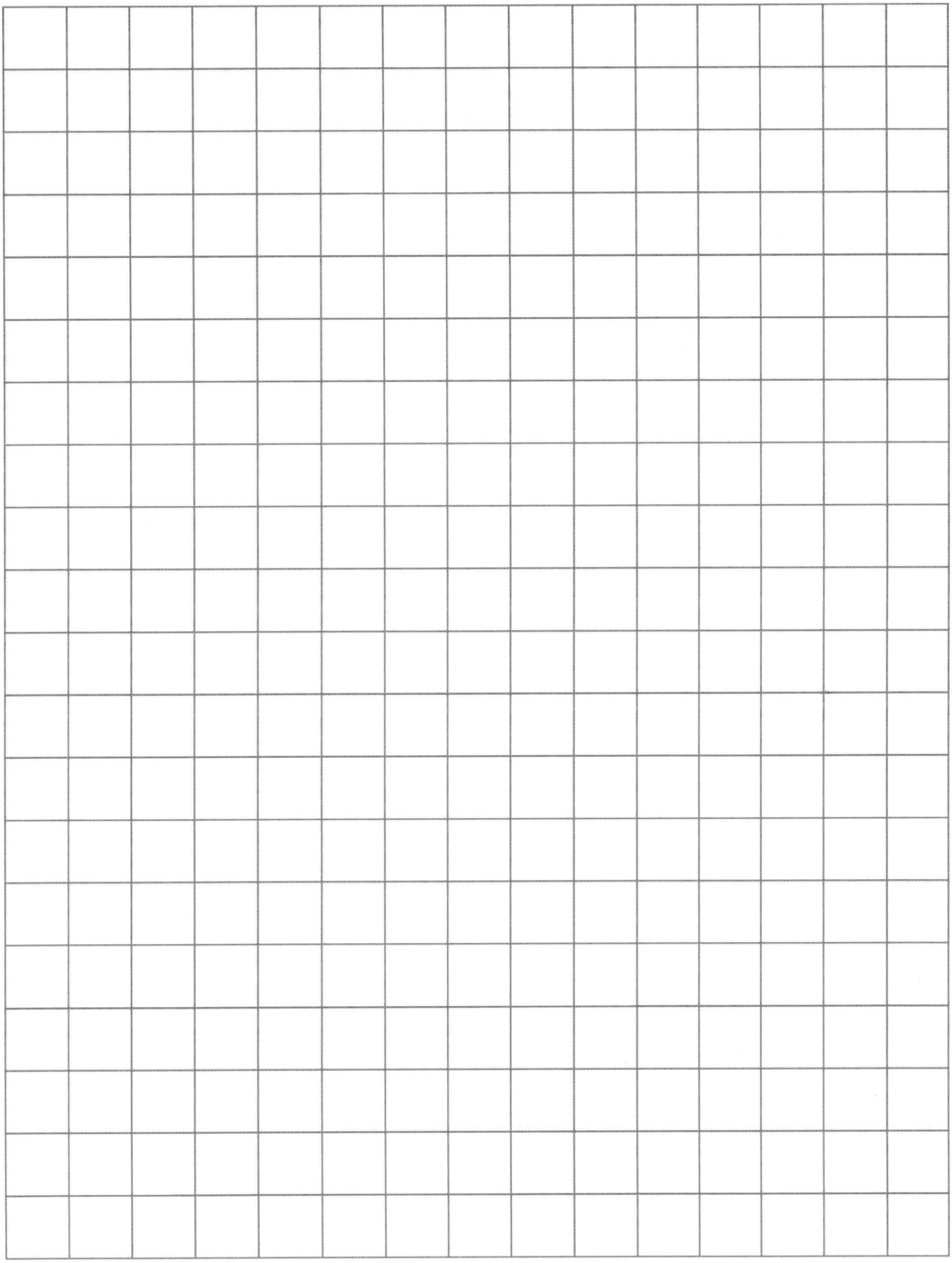

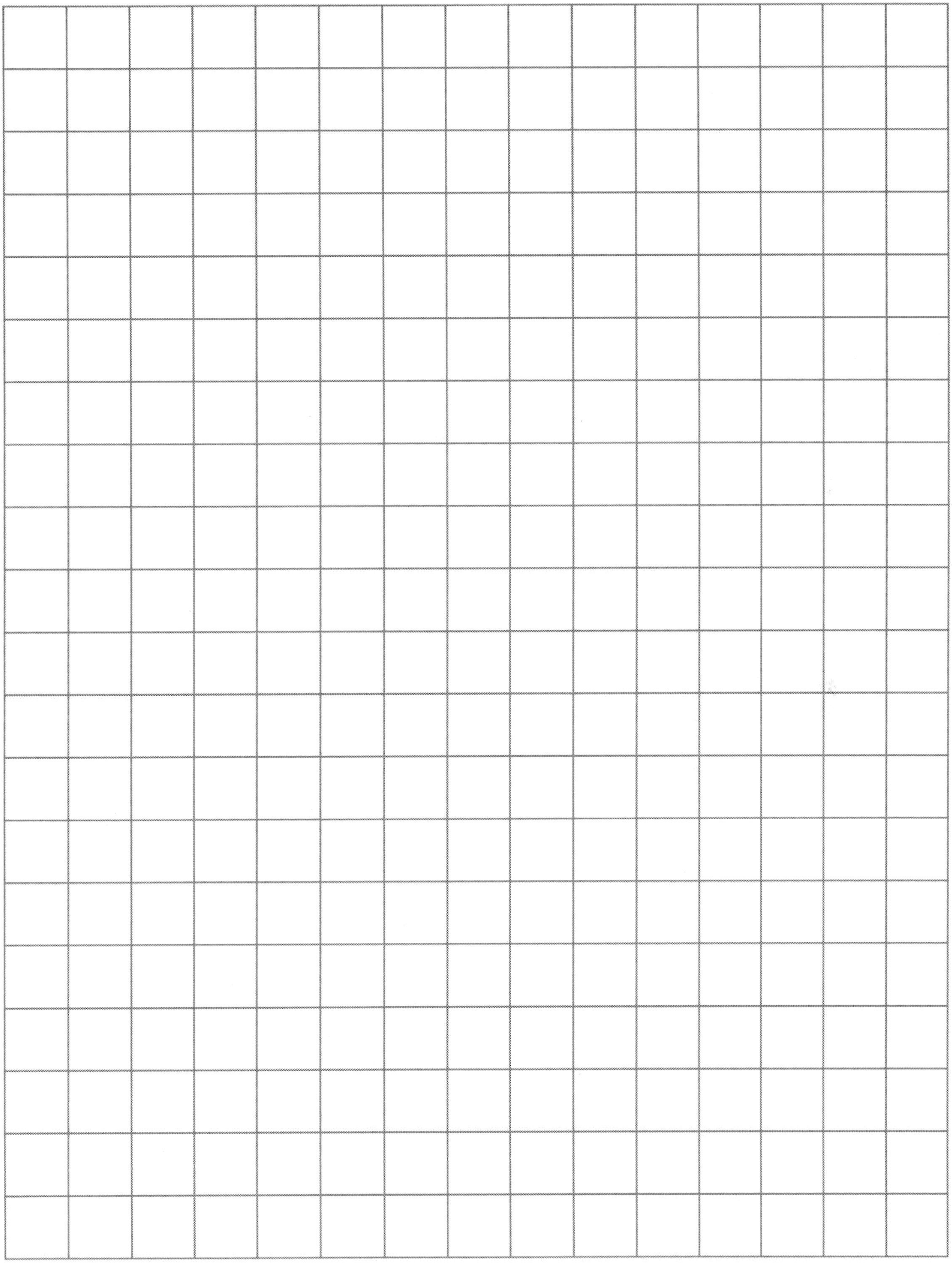

Made in United States
Troutdale, OR
11/03/2026

56845128R00069